SHARKS!

SHARKS!

Tiny Conservationists

VILLAGE LANE
PUBLISHING

Written and Illustrated
by Bethany Heath

ISBN: 978-1-951496-42-5
Village Lane Publishing

VILLAGE LANE
PUBLISHING

For Olivia

There's something
in the ocean
that's really pretty neat!

They have many rows of teeth
and some don't need to sleep!

They smell their food
from far away,
and some see in the dark.

"What could it be?"
you're asking me.
Surprise!
It is a shark!

Sharks are smart,
and fast, and strong,

and sometimes
very stealthy!

We need sharks
because they help
to keep
our oceans healthy!

The whale shark
is the biggest fish
in the ocean blue.

But they only eat plankton
which is much smaller
than you!

Thresher sharks
have little mouths,
and very big round eyes,

and tails like whips
they use to catch
their prey by surprise!

Dwarf lantern sharks are
the smallest
sharks in all the land!

They're so small
that you could even
hold one in your hand!

Hammerhead sharks,
as you may guess,
have funny-looking faces.

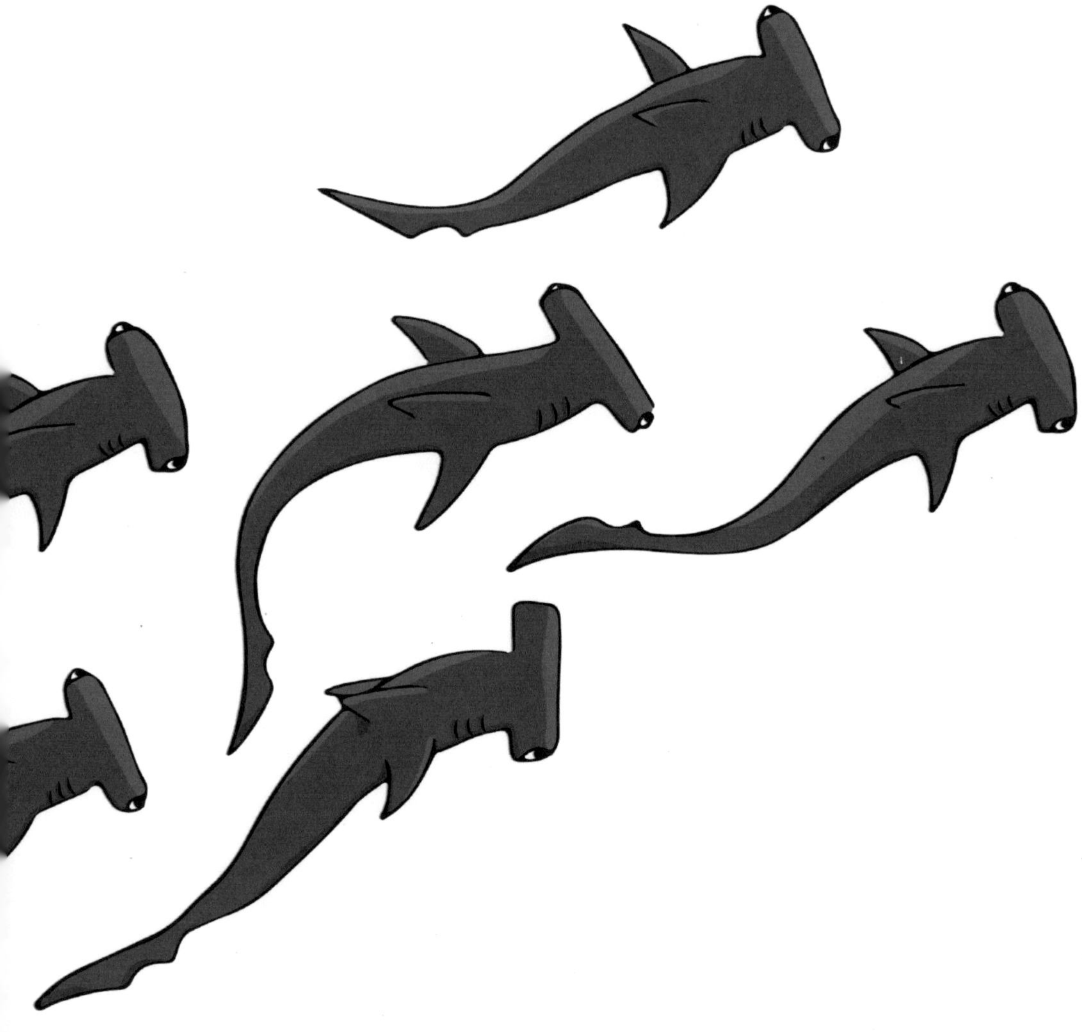

They swim around in groups
called "schools"
when they want to go places.

Nurse sharks are nocturnal, so they sleep during the day.

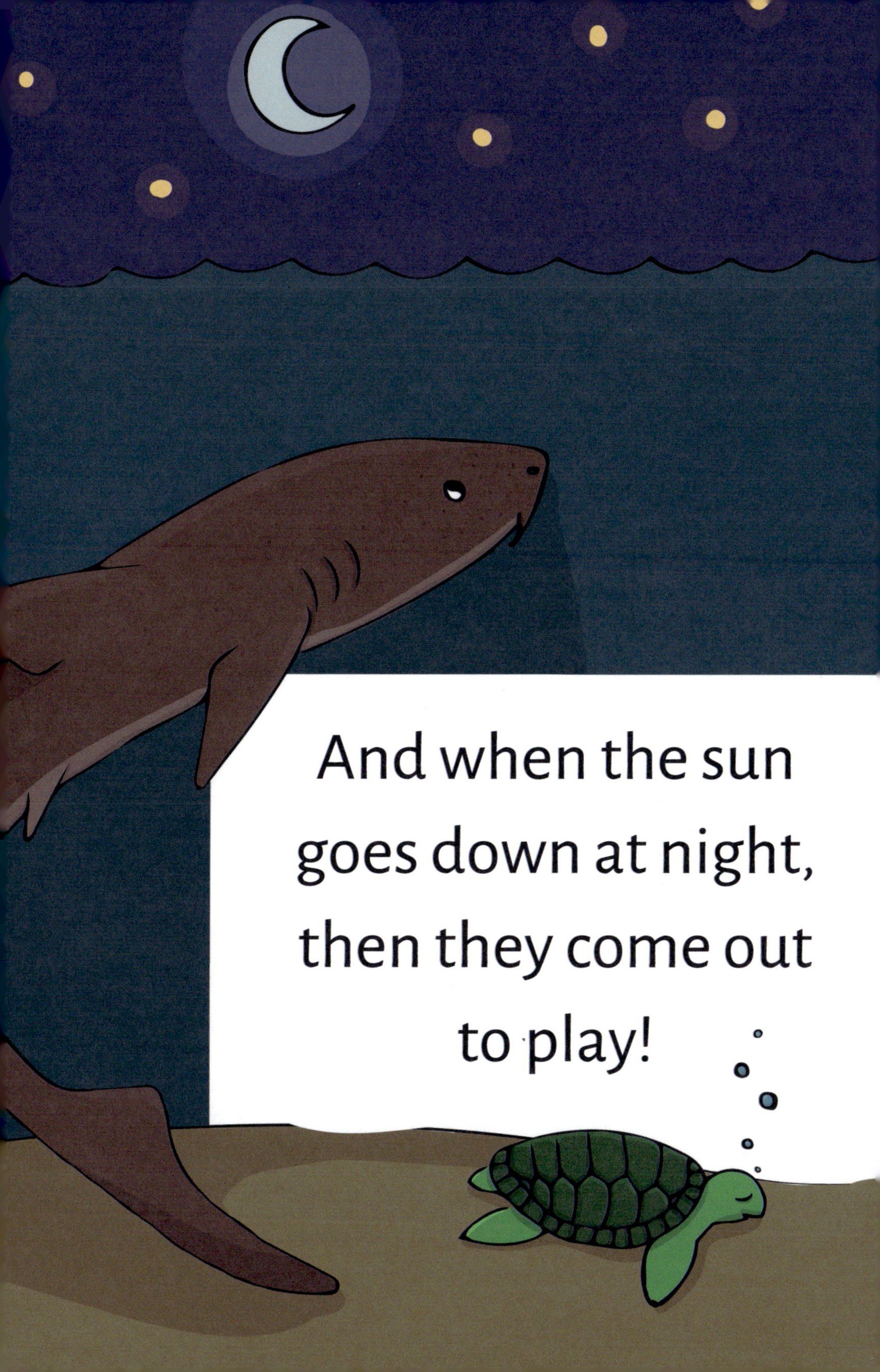

And when the sun goes down at night, then they come out to play!

Mako sharks are very fast!
Their tails have lots of power!

In fact,
these long and speedy
sharks go 40 miles per hour!

Wobbegong is a funny word!
It means shaggy beard.

Wobbegong sharks
are very bendy
and a little weird.

Once there were a lot of sharks swimming in the sea,

but now
there aren't as many sharks
as there used to be.

We need to work together
and do all that we can do!

Because sharks all need a hero, and that hero is you!

How YOU can help sharks!

1. Never litter! Even far away from the ocean, litter is never good. Trash can blow a long way and is very harmful to all kinds of wildlife.
2. Learn about your local recycling laws, and recycle what you can.
3. Choose re-usable products over single-use ones. The more we reuse, the less trash we produce.
4. Learn more about sharks and other ocean life, and share what you learn with others! Knowledge is power!
5. Visit AZA accredited aquariums and ask how you can help.

Thank you for being a conservation hero!

Made in the USA
Monee, IL
05 November 2026

67948396R00021